GW00373014

The Basics of

ELECTRONIC DATA INTERCHANGE

written by

KEITH BLACKER
MSc, CEng, FIEE, FIMgt, FBPICS

illustrated by

ALAN O'COY

EDiSTONE Books

Electronic Data Interchange is a technique used to communicate business and information transactions between the computer systems of different companies and organisations. It was first used in North America in the 1960s and by the mid-1980s it was being used in automotive manufacture, retailing, distribution and holiday booking. Its use is growing quickly and it is set to become the normal method by which many organisations will communicate formally with each other.

This book is one of a series that sets out the meaning and purpose of electronic data interchange. This first title "The Basics of Electronic Data Interchange" is the foundation book for the series and explains the essential features of the technique and gives several examples of its use.

Acknowledgement:
The publisher gratefully acknowledges the assistance of Lucas Industries plc in the preparation of this book. The front cover is produced with the kind permission of The MetroCentre, Gateshead

First Edition 1993, reprinted 1993
Second Edition 1994

ISBN 1 897815 04 2

A catalogue record for this book is available from the British Library

Published by EDiSTONE Books, West Winds, Inkford Brook, Wythall, Birmingham B47 6DA, United Kingdom

Contents

What is Electronic Data Interchange all about?

What is Electronic Data Interchange being used for?

Formal documents that have traditionally travelled from one organisation to another on paper can now, in very many instances, be sent through electronic communication using a technique known as electronic data interchange.

This electronic message exchange is being used in trade, for example, to place orders, to make payments, to exchange product designs and to update catalogues. In the health sector, it is being used to exchange laboratory test results between hospitals and doctors in the community. European police authorities are obtaining driver and vehicle licence details from registration authorities in other Community countries. Teacher pension details are being exchanged between local education authorities and a government department. The list is becoming very extensive.

When electronic data interchange was first developed, the objective was to move existing documents more quickly and more accurately at potentially lower costs. Nothing else was expected to change. However, these very characteristics have encouraged innovative thinking about the way organisations co-operate with each other. Distance could be less of a barrier to trade; the cost of trading over different time zones could fall; it became possible for a supermarket to order and take delivery of fresh goods from the farmer every day. Thus the availability of electronic data interchange is creating opportunities for new patterns of trade, for new administrative communities and different ways of providing and using services.

But let us go right back to the beginning and find out how it all started.

Why did Electronic Data Interchange come about?

In the early days of computers, the sole method of entering data into the machines was by means of punched cards or paper tape. Equally the sole means of getting the results out

An example of a batch computer in the background with its card reader and line printer for output. In the foreground is an on-line terminal for interactive processing

was by means of a line printer. Jobs had to be done one at a time. So the calculation of a manufacturing plan for a factory had to be processed at a different time to the calculations of the wages and salaries. All of the input for any one job had to be entered in together and the total results were printed and distributed at the end to those people that required the information. This is **_Batch processing_** and is still widely used.

Before long, it became possible for the larger computers to process several batch jobs at the same time so making it easier for computer installations to schedule the work according to the needs of its users. This is **_Multi-tasking_** and is the way in

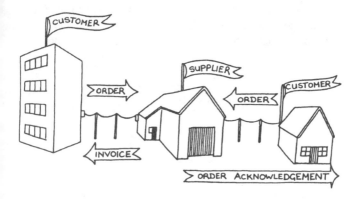

Examples of messages being sent between companies using electronic data interchange

which most mini-computer systems and larger computers operate. Personal computers are also beginning to have this ability.

It also became possible to attach terminals to these machines with each terminal running its own unique job or running a computer program where the user could communicate with the computer to perform a particular task. This is **_Interactive processing_** and is now very common.

Terminals began to communicate with computers at some distance from the building in which they were housed. This makes use of telephone lines with modems attached. Modems are devices which convert between the digital signals used in computers and the analogue signals used in the telephone network. Even satellite and radio connections are used. Computers began to be linked to other computers in the same building or organisation. *Data communications*, as this is known, forms the backbone of modern computing networks.

Taken together, these features of modern computing systems have enabled industrial, commercial, governmental and social organisations to create a wide range of applications such as holiday booking systems, electronic fund transfers and advanced manufacturing. It has even enabled some computers to design computers of the next generation.

Organisations may use a range of computers which are linked together to exchange information between them. This is possible because the organisations have invested in the special equipment and software which is needed to overcome the differences in the way each machine type has been designed to communicate. But the ability to extend this idea to communicating information between organisations has been hindered by these very differences.

Electronic Data Interchange is the technique, based on agreed standards, which enables computers in different organisations to successfully send business or information transactions from one to the other. Business transactions include *orders*, *invoices*, *delivery advices* and *payment instructions*. Information transactions include *income tax returns* and *export statistics*. Traditionally, these are transactions which are exchanged on paper. But Electronic Data Interchange, or EDI as we will now call it, has made it possible to transfer this information directly from one organisation's computer to that in another organisation.

What are the basic components of an EDI link?

For one computer to talk to another there will be some form of data communications link which enables digital data to be passed from one end to the other in either direction. This is the *communications link*.

At each end of this link there is software to set up the data

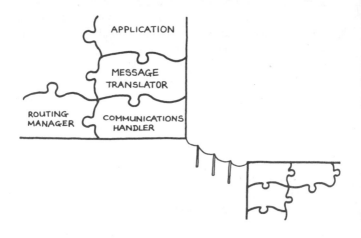

Components of an electronic data interchange link

exchange conversation between the two computers, to look after the sending of the data and to check that no errors occur. This conversation is known as the *communications protocol*. The *communications handler* looks after this communications protocol which has to be the same for both sending and receiving computers, although the software itself can be different. It sets up the call at the beginning of the session and closes it down at the end. The handler also automatically corrects any errors which may occur in transmission.

The *routing manager* records how to set up the communication link for each different partner organisation so that the information can be sent and received. It is similar to a telephone directory giving the number to be dialled to connect to a particular partner.

Connected to the communications protocol software is a *message translator* which takes data which is destined for transmission and converts it into the message standard which has been agreed with the organisation that is to receive the data. Similarly, messages which have been received from the other computer are converted from the message standard into a form of data that the business application can use.

Lastly comes the *computer application* itself which prepares the data to be sent or uses the data which has been received. A stock management application, for example, could send purchase orders requesting replenishment products and a customer order processing system could receive the orders and arrange for the products requested to be packed and sent.

The role of standards

Many organisations set social and technical standards of their own. They may choose which language people should speak in a particular office or whether engineering designs will use metric measures or feet and inches. Equally, they may set standards for the computer languages to be used, for the way data is to be stored and for the communications protocols to be used between their computing machines.

If two organisations need to link their computers, it is likely that each will use different types of computers and will use different communications protocols as well as storing their data in different ways.

Where only two organisations are involved, it is not difficult to reach agreement on which standards to use and work out how to cope with the differences. If a customer has 200 suppliers and each of those suppliers has to deal with 50 other customers, it becomes impractical for each pair of organisations to come to a separate agreement.

Thus standards which are agreed within a trading community

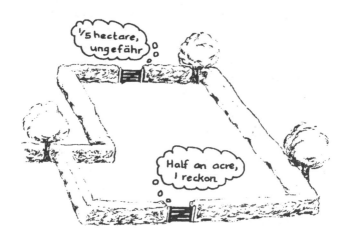

Common standards enable different people from more than one country to understand the same information

such as the retail sector or banking industry can simplify communication within that community. Standards which are national, or international, enable different communities to trade together electronically. The supermarkets can then talk to the banks using those standards. Similarly the road hauliers who move food products to the supermarkets can use one set of standards to communicate both with its high street customers to advise the goods which are being delivered and with its motor trade suppliers for spare parts for the vehicles.

The lack of agreed standards can lead to unfortunate misunderstandings

EDI messages

What is a standard EDI message?

An EDI message collects together various pieces of data which together can be used to perform a particular task. An

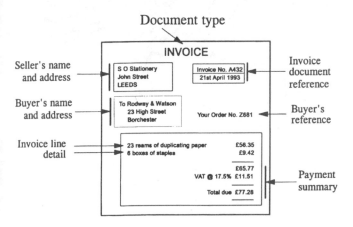

Document type

INVOICE

Seller's name and address → **S O Stationery** / John Street / LEEDS

Invoice document reference ← **Invoice No. A432** / 21st April 1993

Buyer's name and address → **To Rodway & Watson** / 23 High Street / Borchester

Buyer's reference ← Your Order No. Z681

Invoice line detail → 23 reams of duplicating paper £56.35 / 6 boxes of staples £9.42

£65.77 / VAT @ 17.5% £11.51 / Total due £77.28 ← Payment summary

The various parts of a paper invoice which relate to separate segments of an EDI invoice message

example of a message is an order. If one person wants to buy a washing machine by post, an order is sent requesting the particular machine.

The order will contain the name and address of the person placing the order, the name and address of the supplier of the goods, the description of the machine with its part number perhaps and the price quoted in the catalogue.

There may be additional information which is not used on every order. A separate delivery address may be included. Special delivery instructions may be added such as "put it in the shed around the back".

In the paper world, the supplier will receive orders on many different styles of paper, each with its own logo and layout, but all containing the basic ordering information.

An EDI message standard defines the core information which any order requires and sets out where each element of information shall appear. This core information must appear on every order and is classed as *mandatory*.

The standard also defines all the various optional pieces of information that may be included if required and sets out where they shall appear. These additional pieces are described as *conditional* and their use will normally be agreed by a particular trading community or even between individual organisations.

How is a message structured?

Messages are built up from blocks of information which are called *segments*. Each segment contains individual elements of information which are called *data elements*.

An example of a segment in the order message already described is the name and address of the supplier. This groups together information about the supplier. Similarly the name and address of the person placing the order is also contained in a segment.

Such name and address segments will contain *organisation name*, *address line 1*, *address line 2*, *address line 3*, *address line 4*, *post code* and *country*. Each of these pieces of information is a data element.

Each data element will be described in a *directory* for the message standard and, for each entry, there will be a description and its representation. The representation of the *organisation name* could be alphanumeric (it may include either letters or numbers) and be up to 35 characters long. A date is likely to be numeric and six characters long. Computers do not need to know about the dots in between, so 5.1.93 or 5th January 1993 is represented as 930105.

A message standard will also define a syntax. In our written

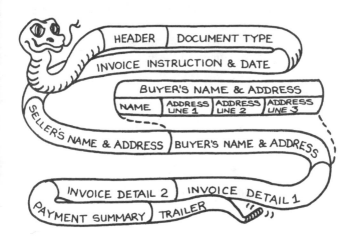

A message is constructed like a snake with a head (called a header), segments and a tail (called a trailer)

language we separate one word from the next by a space, one sentence from the next by a full stop and one paragraph from the next with a spare line and often with an indented start to the first sentence. These conventions are known as the syntax of the language.

In EDI, the syntax enables a message translation program to understand where a message begins and ends, where each segment begins and ends, where each data element begins and ends and marks where conditional elements that are not selected have been omitted. A syntax allows complex messages to be constructed but which are easily interpreted by the receiving computer.

Examples of different messages

In the process of ordering and delivering goods, the list of potential messages between the customer and supplier can be

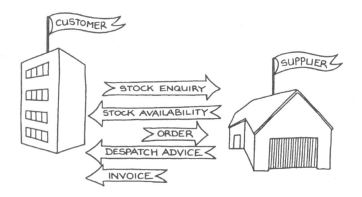

There are different messages used at various times when a customer is purchasing goods from a supplier

quite long. In practice, only a selection of these will be used between a customer and any one of its suppliers.

Examples of messages are *request for quotation*, *quotation*, *order*, *order acknowledgement*, *order amendment*, *order amendment acknowledgement*, *delivery advice*, *receipt*

13

confirmation, *delivery discrepancy advice* and *invoice*.

The process of instructing a bank to make payment and receiving advice from a bank that funds have arrived may use quite different messages.

These could be *payment instruction with remittance details*, *payment instruction acknowledgement*, *credit advice with remittance details* and *credit advice acknowledgement*.

Different types of relationship use different messages.

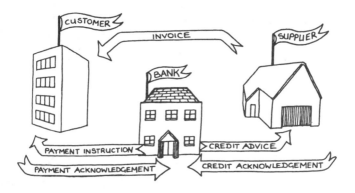

There are other messages used to make payments through the banks

Examples include booking freight on to international shipping, reporting school examination results and declaring customs entries for imported and exported goods. These each have their own special messages.

However, some messages will be shared with other types of relationship. The shipping line will request payment from its customers using the standard invoice message but this will be accompanied by a statement of charges message which is special to shipping.

This additional statement sets out the very detailed information about the loads transported and the ship on which they travelled. This is not included in a normal invoice message.

Different message standards in everyday use

Early message standards were developed by groups of organisations that had a common interest in trading together using EDI.

One example is the airlines which together prepared a method of ordering spare parts from the aircraft and equipment manufacturers. Standard messages for this application were developed through the Air Transportation Association of America.

The North American automotive industry decided that the cost of manufacturing vehicles could be reduced by improving the way in which component parts were ordered by using EDI. The Automotive Industry Action Group was formed to develop suitable standards. The European motor industry set up a similar project to produce its own *ODETTE* standards.

The Article Number Association in the United Kingdom is well known for the standard international bar codes used to identify items in many trade sectors. It too set its own EDI standards - *TRADACOMS* - initially for use between manufacturers and retailers. These were introduced in 1982 and have been adopted by many other sectors such as distributors of electronic components and are used very widely in the UK.

Some standards have been developed at the national or continental level. Here the ones most widely used are those from the American National Standards Institute otherwise known as *ANSI X12* standards.

At the end of 1985, the United Nations recognised the need

to have EDI standards which would support trade between every part of the world and across a wide range of industries and so it created the *UN/EDIFACT* project. This initially involved people from Western and Eastern Europe and North America. It has now become a truly world-wide project with Australia, New Zealand, countries of the Far East, South America and Africa all joining in. The list of participating countries will continue to grow for some time to come.

| TRADACOMS | UK retail & others |

European Automotive | ODETTE |

| ANSI X12 | North America

World airlines | SPEC 2000 |
spares for aircraft

| VDA | German automotive

French retail | GENCOD |

Different trading communities have tended to develop their own
message standards

The project has separate working groups looking at messages for trade, transport, customs, banking, construction, statistics collection, financial services and the health sector. It is likely that new working groups will be created as more types of businesses and organisations find that EDI will help to improve the way they work.

The chemical, electronics and construction industries in Europe pioneered the first UN/EDIFACT messages - just the ORDER and INVOICE to begin with.

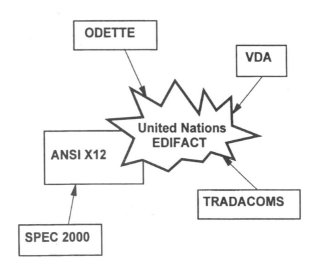

Many local standards are now migrating to the United Nations EDIFACT standard

It can take two to three years to introduce any international standard because many people wish to get involved in its creation . This takes a lot of time and money and many of those involved are volunteers. UN/EDIFACT messages are beginning to be approved in large numbers and they are all being adopted by a larger number of industries.

It takes time for everybody to change over from using standards that they are familiar with to a new set of international standards. By the year 2000, much EDI will be based on UN/EDIFACT but it is likely that there will still be considerable use of the older standards.

What is the purpose of a message translator?

Part of the specification of a message standard is the *syntax* which it uses.

The early message specifications used the punched card as a reference so that the first so many characters had a particular

Message translation is like wrapping up a parcel at the start of a journey and unwrapping it at the end when it is safely delivered

meaning and the next specified number another meaning and so on. As the need arose to design more complex messages, more sophisticated approaches to message syntax were required.

When a computer application is preparing to send information via EDI, its creates the data required by the message but in a form which is normal for that particular computer. The *message translator* takes the data in that form, checks that the information items are correct for the message standard and then adds the syntax.

When receiving information, the message translator removes the syntax and checks that all the information is presented according to the message standard.

It is somewhat like wrapping up and addressing a parcel with a number of different things inside before sending it on its journey through the post. To get to its destination safely, the wrapping must be robust enough for the items to be transported without damage and the address must clearly show the destination so that the delivery service can properly route it through its network. On arrival, the wrapping is taken off and the items unpacked and used for their intended purpose.

Types of communication link used by EDI

Value added networks

Much of the electronic traffic between companies does not have to arrive as soon as it is sent and, in any case, a smaller company's computer may not be ready to receive information at the time that the sending company wishes to pass the information on.

Large communications networks have been set up to provide post box - mail box services between EDI users. These networks are known as *value added networks* (VANs) or *clearing centres* and are often built and run by the public telephone companies or large computer suppliers.

A company which wishes to send EDI messages dials up the VAN and deposits packages of messages into its post box , each package of which includes the electronic address of the recipient. The VAN takes the data from the post box and sorts it into the recipients mail boxes ready for those users to collect when they each dial in next. Thus at the VAN, each EDI user has its own post box into which it puts outgoing messages and its own mail box from which it retrieves incoming messages.

These VANs provide other basic services such as message tracking and the ability to link types of computer that could not otherwise link directly. By tracking messages through the network, VANs record whether and when messages arrive, are transferred and are picked up. So if there is a fault at any point along the end to end link or there is a dispute between the users, evidence can be produced to show what really did happen. This is similar to the registered letter service provided by the postal services.

VANs have developed the capability to exchange data with a

wide variety of different computers using the appropriate communication protocols. As long as any two users can individually link to the chosen VAN , they will be able to send messages to each other through the VAN even though they may not be able to link directly to each other.

There are many different VANs around the world, some

A Value Added Network provides an easy means of accessing many other users and it also correctly sorts the messages in transit like a postal service

extending to many countries, others providing national coverage and yet others which serve only a particular industry or community. In the early days of EDI, a user was obliged to subscribe to as many different VANs as its partners (usually customers) required. VANs are beginning to interconnect so gradually obviating this need to subscribe to more than one network.

Direct connections

Some large companies operate their computers most hours of most days and so have decided that they will not use the value added networks. Instead they have chosen to link directly to their customers and suppliers. These *direct links* can use normal telephone lines with modems at each end or special links which have been designed to carry digital data.

Direct links cost less than the use of another service in the

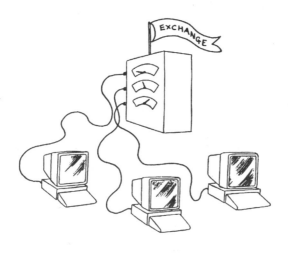

Communicating data directly between one user and another through the public telephone network

middle but they do have the disadvantage that an organisation may have to make many network connections when placing orders with all of its suppliers. If it used a value added data service it may have needed just one connection to send all of the data to all of its suppliers. The service in the middle then behaves like an electronic postal service where letters for lots of addresses can be put into a post box and the mail service will sort them and send them to the right destination.

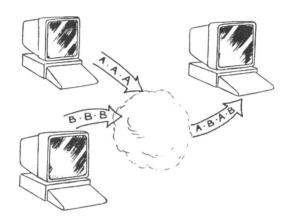

A packet switching service is more like a postal service by telephone where a single delivery from the postman can contain letters from a variety of different people

An intermediate form of link is called *packet switching*. There is a single link between the each company and a packet exchange network provided by the telephone companies. Any one terminal then sends information for many different destinations down that one line all interleaved together. The exchange network sorts out the route for each packet and sends it to the correct location. There is no mailbox in this particular option which means that the receiving terminal has to be listening and operating at the same time as the sender is transmitting information to it.

Immediate response communications

In some situations data has to be processed soon after it has been received and a reply quickly returned to the sender. When a manufacturer is asking suppliers to provide components just-in-time, a requesting message may require a reply within a minute or two. Just-in-time enables a manufacturer to eliminate stock holding by calling off material from a supplier at very short notice. It is usual then for the manufacturer to use a direct link to the supplier or to use a network which has been designed especially to give a fast response time.

In other situations, a person working at a terminal may need direct access to an application in another company and will want immediate response to questions asked or requests made. When booking a holiday or ordering a motor spare, the customer will often be adjacent to the salesperson and expecting an immediate answer about the availability of an airline flight or an exhaust pipe. In these two cases, the communications link must be designed to give the speed of response that the customer is expecting.

The form of the EDI message may be the same whether an immediate answer is required or whether tomorrow will do. The design of the communications link and the computer application at each end will depend on the speed of service that is required for each type of user.

Can these communications be secure?

There is sometimes a fear that electronic communication is not quite as certain as using paper. If you do not have a paper document any more, can you be certain that the electronic record will not get lost? If the EDI message contains personal information such as a medical record, how can you be certain that it will not get into the wrong hands or be changed in some way during transmission?

And yet we are familiar with cash machines at banks and credit card terminals at cash desks in stores. Both types of machine are linked securely to our banks, our identities are authenticated and the correct amounts of money debited from our accounts.

All the value added networks offer a sensible level of security and authentication. Security is provided in the design and quality of the service they operate. Authentication is provided by the use of identifiers and passwords when a user makes a connection. This level of control is sufficient for most applications. A similar type of authentication can be used for direct connections.

A higher level of authentication can be provided by the use of a smart card. In these circumstances, the smart card has to be read in during the sign on cycle when the terminal connects directly to the destination or to the VAN.

If very secure EDI communications are required, encryption can be used so that the receiver can be sure who the sender is, the sender can be certain who the receiver is and both know that the transmission has not been interfered with on the way. Suitable techniques are already used in defence communication and these can be used for EDI links where this is considered to be necessary.

Hardware and software options for EDI

Using a Personal Computer as an EDI terminal

There are various ways in which computers can be set up in order to send and receive EDI messages. Firstly, a Personal

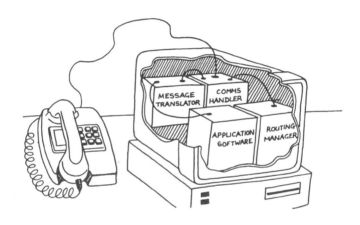

The types of software packages which would make up an EDI terminal on a Personal Computer

Computer can be used independently as the company's link to the outside world. Alternatively, a small computer can act as the interface between the outside world and a separate computer which processes the business applications. The third option is to link the applications computer directly to the outside world. The option chosen usually depends on the scale of EDI use and its importance to the organisation's operation.

But using a Personal Computer with a modem and communications line is an easy way to start using EDI. Software is available which provides all the necessary EDI functions such as the communications protocol, and the EDI message translator.

The role of the routing manager may be included in the software so that each communication link is set up automatically each time a data exchange is required. Alternatively, it is quite possible with a personal computer to establish the calls by manual dialling through the telephone hand set.

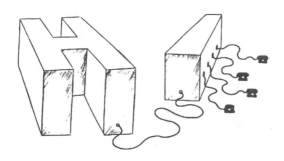

One computer looks after the EDI communications while another looks after the applications which create the information to send or use the information which is received

Off the shelf software packages are available which provide these EDI functions and offer the ability to enter the details of a message for sending as well as the ability to display any messages that are received. In addition, it is possible to print the messages on paper so that the information can be handled as if it had been received on paper.

28

Using a small computer as an EDI front end processor

Some organisations have one large capacity computer or a network of smaller computers to run their various applications such as design, stock management, order processing, statistics management or accounting. The EDI processes can be run on a separate front end computer which is linked to the larger machine or the network. This front end computer undertakes the EDI communications, the message translation and the routing management.

The computer on which the application runs exchanges the raw information with the front end computer which in turn looks after the various EDI and communications tasks.

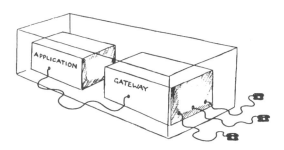

The one computer is looking after both the EDI communications and the applications which send and receive the information

An EDI gateway on the applications computer

The EDI processes can be run on the same computer as the business applications that receive or send the information.

If these processes are grouped together and provide a common EDI facility for a range of different applications, this is known as a gateway. Such a gateway usually has additional facilities such as keeping a record of all the information which is received into and sent from a company via EDI. It also manages the overall routing to the various partner organisations for every business application.

A single gateway for larger organisations

A company and organisation may use just one gateway to link to the outside world for all its applications. These applications can be running on the same computer as the EDI gateway itself or may be networked to the gateway from other computers on the same site or on different sites which could be far away.

Such a single gateway enables the organisation to present a common image to its customers, suppliers and partners for its electronic communications even though it may conveniently operate internally as a number of smaller separate units.

An organisation may also choose this style of connection because it wishes to have just one gateway to manage rather than many on different computers and on different sites.

Interfacing EDI into an organisation

Integrated EDI communications

When integrating a computer application in one company with another application in a different company, it is necessary to ensure that the information which is transferred means the

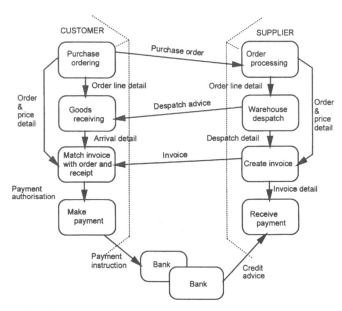

The EDI communication is directly between the computer applications in one company and those in another

same thing to both the sending and receiving organisations. For example, if a product which is being ordered is specified by a part number, there needs to be a decision on whose part number will be used.

Manufacturing companies usually have their own sets of part numbers even for components that are purchased from suppliers. So, in addition to agreeing whether the ordering part number will be that of the customer or of the supplier, it is also important to agree how the part number catalogues in the two organisations are to be kept up to date.

When the delivery date is requested, the choice of date has to be agreed. It could be a specific date, not before the quoted date, no later than the date requested, or sometime in the week that includes that date.

Some items of information are quite obvious from the message specification, but other parts of the specification are open to interpretation and need to be mutually agreed if the meaning of the information is to be transferred successfully.

It is also necessary to agree the meaning of coded information. Some codes, such as currency - GBP for Great Britain Pounds (£) and USD for United States Dollars - are agreed internationally by the United Nations. Other codes may be very local but everybody using them must know what they mean.

When processing EDI messages automatically, it is useful to recognise that a message is built up from segments rather as a toy model can be built from bricks. Processing a received message can be viewed as taking the model apart and doing different things with each brick. For an order, the sender name and address segment can be used to find that organisation's record. Each order line segment can be used to check stock availability and create a picking note. The delivery address segment can be used to create the delivery instructions.

This idea can be extended to building new messages out of old ones. An order message received can be converted into an invoice by copying the name and address segments and order line segments and then adding further segments giving price details, tax details and the totals due.

The link with bar coded labels

Many industries are using bar coded labels on goods or packages. These can be used to identify what the item is. For

example, the bar code on the back of this book describes the International Standard Book Number for this title from EDiSTONE Books. Such labels can be read at a book shop cash desk to record its sale. A package label may either describe a code for the contents or a code for that particular shipment. These codes are reference numbers and other information is needed to interpret them either into a product description - "12 447gram tins of Sainsbury's baked beans" - or into the contents of the particular shipment - "12 radio sets despatched on 12th October against order number 234/A22".

EDI messages, sent in parallel with the delivery of such bar coded packages or products, can provide the interpretation of these codes. Each bar coded reference number can be presented in the message alongside the particular details that the recipient requires, whether it be product descriptions, unit price, order details or any other such information.

If the EDI message arrives ahead of the packages, the bar coded labels can be read by a computer that can immediately associate the reference numbers with the details required for checking the contents or for routing the package to its destination.

EDI can affect the way an organisation works

When orders that formerly arrived on paper begin to appear electronically, some organisation start to look at the whole order processing operation to see how it can be improved. The task of order vetting that was previously manual could be carried out automatically. The orders could be acknowledged electronically so increasing the customer's confidence in the order being delivered as expected. Sometimes the whole order processing cycle will be examined with the aim of reducing the order-to-delivery time or to make the response time to orders more consistent.

All of the examples just quoted involve orders, but there are many similar instances in, for example, making payment, in hospital care or in airline operations where the potential of EDI is acting as a catalyst for basic operational improvements. These improvements can be individual changes such as the introduction of the acknowledgement or they can form part of a wider integrated programme of change, often known as "business process re-engineering".

Such re-engineering is typically associated with a particular business or operational objective. If the target is to reduce stockholding by 50%, a whole new commercial framework with suppliers may be required including shorter supply lead times, more frequent deliveries as well as smarter communications using EDI. If a hospital wishes to reduce its waiting list to a maximum of three months, changes may be required to the way beds are scheduled, operating theatre time is allocated, doctors' diaries are managed as well as changes to communications with community doctors and suppliers of medical supplies and drugs.

Sometimes the availability of EDI is the trigger for change. In organisations where the understanding of EDI is fairly mature, it is more likely that a strategic requirement will drive a programme of business process re-engineering that will, in turn, require EDI communications with other organisations.

Comparing EDI with other forms of communication

How does EDI compare with the use of paper documents?

Paper has been the principal means by which business has been carried out for many years. Most people are familiar with documents on paper especially when they are signed to make them official.

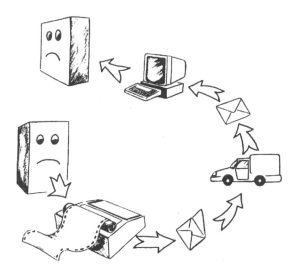

In commerce and industry, much of the input to one computer is the output from another. The link is usually on paper through the postal service

If paper documents are sent through the post, it takes time. The delivery of documents by hand is sometimes quicker but this can be very expensive. The cost of preparing documents, putting them into envelopes, posting them, unwrapping them and probably entering them into a computer will get even more expensive over the years as the cost of data communications is falling while the costs of postage and people's time continues to rise.

For many organisations, most of the information arriving for

EDI can be much tidier, quicker and more accurate than working with paper documents

entry into a computer has been prepared and printed on somebody else's computer. It is very easy to make a mistake when reading information from paper documents and keying it into a terminal. Documents can stick together, the wrong key can be pressed accidentally. If the letter or figure is difficult to read, it may be interpreted incorrectly.

The use of electronic data interchange does avoid the need for a person to enter the information again into another computer and it does eliminate the possibility of making those human errors.

The real problem of making mistakes is that the wrong

goods may be delivered or the package may be sent to the wrong address. This not only causes additional costs in putting the mistake right, but more importantly it makes the quality of the service look poor.

How does EDI compare with the use of the telephone?

Telephone ordering or gathering opinions and information by telephone is quite common.

As with the use of paper documents, a person at the receiving end, has to key the information into a computer and

Telephone answering uses a person at each end to interpret the information being sent by one to the other

it is possible to make similar mistakes, not only in the entry of the information, but also in interpreting the information given by the caller. If the caller and receiver have different native languages, the prospect of making a mistake is even greater.

Telephone answering does enable the two people in different organisations to get to know each other which can provide the important personal touch but there is the risk of reducing the quality of the overall service.

How does EDI compare with other forms of electronic communication?

It is possible to use electronic mail to transfer information from one company to another. Such a message could represent an order, or a delivery instruction, or an examination result.

Electronic mail is particularly oriented towards communication between individual people and does not require the sender to put the information into a particular sequence or

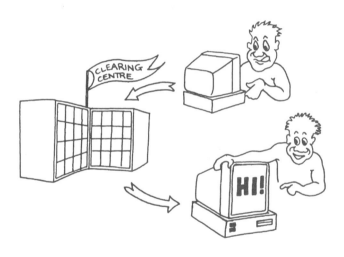

Electronic mail is essentially communication between people - not computer applications

into predetermined positions on the screen. If the receiver is another person who can visually interpret the instructions, the information can be successfully processed by hand. However, if the receiver is a computer application, unstructured information cannot be understood or processed automatically.

Electronic communication can be used to link two computers together to send messages or files of messages to a locally agreed design. Such communication can be quite successful, but it does require the pairs of organisations to define and agree what information is to be sent.

If one organisation wishes to change the message or file design, then the other parties have to change their computer programs to match. EDI is distinguished from this type of transfer by the use of national or international standards of message design and using generally accepted standards for communications protocols.

Facsimile transmission also makes use of data communication between two particular stations. This is used for document transfer which may originate from a machine reading a piece of paper or from a computer system itself. The destination is usually a printer at the receiving station. What that receiving station is getting is a series of dots which, to the human eye, make up a picture or a series of words.

Like electronic mail, facsimile transmissions do not allow a receiving computer to interpret any message transferred so that it can be processed automatically. Although such transmissions are just as quick as EDI transmissions, there are extra steps involved such as going to collect them from the facsimile machine and entering them into the computer. These additional steps take time and can give rise to errors just as with the paper and telephone communications.

Electronic mail, facsimile transfer and message or file transfer each have their own place, but they are separate from EDI with its role of transferring formally structured information for automatic processing.

Some examples of the way EDI is being used

A supermarket was anxious to reduce the number of times it ran out of stock on its shelves while, at the same time, holding smaller quantities of those particular items in its stock rooms and warehouses. Electronic communication is used to collect the stock levels at the stores every day and these are used to calculate the orders for the dry goods, frozen goods and fresh items which are transmitted to the suppliers by EDI.

That same supermarket was finding that the cost of receiving

In the United Kingdom, the well known names in the High Street are big users of EDI for placing orders for goods on the supermarket shelves

and processing the large number of paper invoices from suppliers was becoming very high. Now it receives EDI invoices which are processed automatically at a much lower cost.

Supermarkets are now beginning to share with suppliers the

market information which is obtained from check-out tills which directly record what people purchase. This helps suppliers who understand how products are bought and also helps in forecasting future demand.

Clothing retailers are also sending orders to their suppliers via EDI to replenish stock that has been sold and to initialise new product lines. This avoids the need to hold very much stock while maintaining customer choice. Some suppliers are required to give their retail customers electronic visibility of the stock held so that the shop knows what it is able to order.

There is a bewildering variety of spares that are required for a range of motor vehicles and EDI is being used to order and quickly obtain the necessary parts

Such close relationships between the High Street clothing retailers and textile manufacturers mean that fashions can be changed very quickly according to customer taste or the nature of the weather.

There is a bewildering variety of spares that are required to service a range of motor vehicles and it is not practical for any retailer or wholesaler to keep all of them in stock. It is very costly and time consuming if, whenever a spare part is required, it is necessary to make a series of telephone calls to find out who has that particular part in stock. One supplier to the motor trade uses EDI to enable its wholesalers to enquire whether a particular part number is available and what its price is. Then, if the answers are acceptable, the wholesaler places an order for that part and receives a confirmed acknowledgement. This total "conversation" takes place between the computer systems of the customer and supplier.

One car component maker is sending EDI invoices to a vehicle manufacturer in much the same way as those being sent to the supermarket. In this example, the return payment is also being made using EDI. The payment instruction and details of the invoices being paid are sent to the vehicle manufacturer's bank. Funds are transferred to the component maker's bank together with the details of the invoices involved. The receiving bank then sends an EDI message to the component maker advising that the funds have been received from the vehicle manufacturer and giving those invoice details. The component maker then updates the accounts automatically.

In just-in-time supply, rapid and accurate communication is essential. One vehicle manufacturer uses EDI to keep its assembly track supplied on an hour by hour basis from an independent warehouse about a kilometre away.

The stock level in that warehouse is maintained by the suppliers who receive daily EDI transmissions detailing what stock has been removed to the assembly line and what has been received from the suppliers' manufacturing plants. The vehicle manufacturer also communicates directly with the suppliers through the use of EDI confirming what has been received at the assembly line, advising of any errors in the type, quality or number of parts received and sending self-billed invoices. Self-billed invoices are calculated by the customer who knows the contract price and the quantity. These are then sent to the supplier to advise what the payment will be in due course.

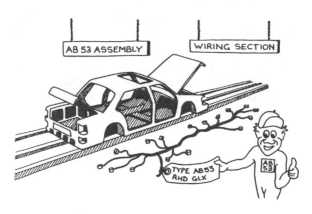

A car manufacturer uses EDI to send orders to a component manufacturer for items to be made and delivered within ten hours

EDI is being used to book cargo space on ships and airlines. Provisional bookings are made to the shipping agents who confirm whether space is available on a particular ship or aircraft. The requirement for that space is subsequently confirmed by the organisation which is exporting the goods.

Container ships carry large numbers of container boxes each of which is destined for a particular port. Its position on the ship can be determined by its length, its weight and whether it needs refrigeration services. A port which is about to receive a ship needs to know where the containers that it needs to unload are located and where there are spaces into which outgoing containers can be stowed. EDI is being used to send the container bay plan from the dispatching port to the receiving port so that the activities of the container terminal can be planned and the ship can be turned around in the shortest possible time.

Customs authorities have a duty to check goods coming into

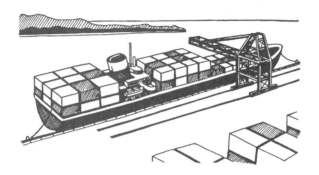

Container lines, ports and customs use EDI to speed the movement of goods and to simplify the customs declarations

the country to trap illegal imports. They also have to identify goods on which excise duty is to be paid and to collect statistics for the calculation of trade balances.

Customs authorities in the UK, as in a number of other countries, accept customs declarations via EDI. This allows them easily to check the entries for correctness, to check for any import licence requirements, to calculate any duty payable and to clear the goods for movement from the port. Some customs authorities even use EDI to clear customs entries before the goods arrive into the country by receiving an EDI import declaration from the exporting company in the country of origin. Thus the only tasks which remain when the ship docks is to check for illegal goods movements and false declarations before the goods are transported off to their intended destination.

In the consumer goods market, warranty claims are claims to replace items of equipment which have become faulty within the guarantee period following purchase. Claims are also a source of information about in-service product failures which can be used to improve the design or method of manufacture. One organisation is receiving warranty claims via EDI from a national distributor in an overseas territory. It is using the information to refund the distributor's costs of replacing the item and to analyse the causes of the failure so that the design and manufacturing processes can be improved.

These are individual examples of the use of EDI. Other examples can be found in similar organisations all around the world. There are very many other uses of communication with between companies, government departments, local agencies and service organisations that have been developed by using the EDI technique.

When compared with other methods of communicating, organisations are finding that electronic data interchange, when applied correctly, does reduce operating costs. Others have found that it enables them to give a faster service to customers without increasing costs. Some use EDI along with other techniques to develop new markets that they could not otherwise reach. New information products are emerging which are EDI based.

What is certain is that there are many more exciting stories yet to emerge. Many of these may well use Electronic Data Interchange in new and imaginative ways . . . . .

INDEX